Entrepreneurship in Chemical Engineering 2019 – Sponsored by CIEE

Topical Conference at the 2019 AIChE Annual Meeting

Orlando, Florida, USA
10-15 November 2019

ISBN: 978-1-7138-0552-6

Printed from e-media with permission by:

Curran Associates, Inc.
57 Morehouse Lane
Red Hook, NY 12571

Some format issues inherent in the e-media version may also appear in this print version.

Printed with permission by Curran Associates, Inc. (2020)

For permission requests, please contact AIChE
at the address below.

AIChE
120 Wall Street, FL 23
New York, NY 10005-4020

Phone: (800) 242-4363
Fax: (203) 775-5177

www.aiche.org

Additional copies of this publication are available from:

Curran Associates, Inc.
57 Morehouse Lane
Red Hook, NY 12571 USA
Phone: 845-758-0400
Fax: 845-758-2633
Email: curran@proceedings.com
Web: www.proceedings.com

TABLE OF CONTENTS

578382 Asserting and Protecting IP Rights in Chemical and Pharmaceutical Industries

Obtaining and defending technological innovations is a critical issue for scientists, academics, and executives to understand. Due to developments in technology and in law, the interplay between different types of intellectual property ("IP") rights are evolving, and not always mutually exclusive as traditionally viewed. I will discuss practical approaches for different forms of IP, using real-world examples and cases, with particular focus on patents and trade secrets. We will discuss recent developments, such as the trade secret cases brought in Federal Court under the 2016 Federal trade secret statute, updates to patent-eligible subject matter, and cases involving multiple forms of IP -- and how they may affect the chemical and pharmaceutical industries.

568355 Patents Are Integral to Monetizing Innovation. Court Decisions Have Led to Uncertainty As to What Is Patentable, Affecting Drug, Diagnostic, Genetic, and Other Innovation

It has been nearly five years since the U.S. Supreme Court issued the *Alice Corp. Pty. Ltd. v. CLS Bank Intern.* decision and about seven years since its decision in *Mayo Collaborative Servs. v. Prometheus Labs., Inc.*. Those decisions established the current framework—dubbed the "*Alice*/*Mayo* test"—for determining the subject matter (*i.e.*, technology) eligible for patent protection. Since the creation of the *Alice*/*Mayo* test, the primary appellate court that handles patent appeals has issued many decisions hinging on this issue, but significant uncertainty remains. This topic is important for those in academia and industry because patents are prevalent and highly-sought after in both arenas.

The *Mayo*/*Alice* test is a two-step analysis used to interpret § 101 of the Patent Act, which states that a patent can be granted for "any new and useful process, machine, manufacture, or composition of matter." The Supreme Court has created certain exceptions to that rule that include natural products, natural phenomena, and abstract ideas. The Court justified the judicial creation of those exceptions by expressing concern that, without them, a patent could cover the "building blocks" of nature and tie-up their use for future innovation. This is complicated, however, by the fact that all inventions are governed by the natural laws and scientific confines of the universe. When an innovator is working to create new technologies that use or rely upon naturally-occurring materials like DNA, proteins, antibodies, and other chemical constituents, it can be difficult to ascertain whether a court might consider a discovery to fall into one of the exceptions and be ineligible for patent protection.

In general, the *Mayo*/*Alice* test implicates patents that rely on computers (often characterized as "abstract ideas") or that incorporate or rely on naturally-occurring substances. The latter can be particularly problematic for those working in chemical and biological fields. For example, the *Mayo* case involved a therapeutic method for treating an immune-mediated gastrointestinal disorder that included the steps of administering a drug and then determining the levels of 6-thioguanine after that administration. The Supreme Court declared that therapeutic method was ineligible for patent protection.

Many important technologies have been implicated by this relatively recent jurisprudence. For instance, patent claims for isolated DNA, a method of assessing the risk of a major cardiac event using enzyme levels, a method for detecting mycobacterium tuberculosis, a method of DNA intron sequence analysis, a method of diagnosing a neurological disorder based on the presence of certain autoantibodies, and a prenatal test for certain fetal characteristics using cffDNA were all found to be patent-ineligible natural products or natural phenomena. However, patent claims for treating a patient with schizophrenia, compositions used for increasing athletic performance using beta-alanine, and a process for cryogenically freezing hepatocytes were found patent-eligible. Despite some common technological threads running through these patents, the courts' opinions have made it difficult to cogently predict how similar patents might be treated in the future.

Logically, overbroad patents are often considered to have an inhibitory effect on innovation because they can tie-up later innovators' ability to use preexisting technology to expand upon it and create new technology using it. Much innovation is iterative, after all. However, the U.S. patent system is designed to encourage invention and innovation and when innovators lose the incentive to invest in and create valuable and complex technology because they have no way to protect those inventions, this can be detrimental to innovation and society.

The author of this abstract recently represented a dietary supplement company in a successful appellate reversal of a lower court's ruling that the supplements at issue and methods of treatment using them were patent-ineligible. The patents involved in that case concerned a dietary supplement containing unnatural amounts of beta-alanine that were administered over an extended period of time, resulting in an increase in the anaerobic working capacity of a subject's muscles. That particular method of administration caused an unnatural override of the body's homeostasis. After the lower court concluded those treatments were nothing more than a natural law, the appellate court reversed that decision and held that the patents sufficiently distinguished themselves from a natural law or natural phenomenon and were patent-eligible. Based on this experience, the author intends to discuss the state of patent law, with particular focus on this issue, along with the courts' treatment of the two-step *Mayo/Alice* test. The presentation would also discuss how innovators may seek patent protection on the types of inventions that a court or the U.S. Patent Office might consider subject-matter ineligible.

583619 Start-up, Spin-out, Don't Burn-Out: Protecting Your Intellectual Assets on a Shoestring Budget

The discussion will include a general overview of intellectual property rights and why they are important for emerging growth companies (e.g., entrepreneurs, start-ups, university spin-outs) to consider early in the life of the company. The discussion will compare and contrast patent, trademark, copyright, and trade secret protections. The discussion will include steps an emerging growth company can perform on their own without having to involve an outside consultant (e.g., attorney). The discussion will include common pitfalls when obtaining intellectual property rights and creative ways emerging growth companies can enforce their rights to deter competition or generate revenue. And of course, throughout the discussion, we will discuss how to pay for it all. This discussion is geared toward non-patent professionals, such as engineers, scientists, managers, and c-suite executives.

587156 How to Protect Your Intellectual Property in the Academic World

Universities are full of academics working on cutting edge technology, but how is all that innovation protected? Universities typically have their own organizations in place to help encourage, protect, and license innovations occurring at the school. This presentation will provide an overview of the basics of patent law, how tech transfer works at a university, and important considerations for students, professors, and the university itself. Speakers will also highlight several case studies addressing whether a university's intellectual property protection was sufficient.

588860 Life Cycle Management for Life Science Inventions

The presentation will focus on strategies for building patent portfolio with timing considerations, diversifying patents with alternatives to standard utility filings, and alerting FDA with a citizen petition on a competing product; to extend market exclusivity for a drug or medical device franchise.

560807 Investing in University Startups: How to Pick the Winners

When most people think of university students starting companies, they imagine web-apps and social networks. On the contrary, the university is a fertile breeding ground for companies across all industries. In the chemical space, exciting university startups are making breakthroughs in fields such as clean tech, pharmaceuticals, agriculture, and beyond!

In this talk, I use my experience as a Partner at Dorm Room Fund to provide a case-study based investigation of the do's and don'ts for university-based startups in the chemical space. Dorm Room Fund invests exclusively in student founders, and in the past 3 years we have invested in 8 companies in the chemical space. I will provide an overview of what made these companies successful, and some cautionary tales about what not to do. I will also provide further insight from a survey of 10 later-stage chemical companies that originated within a university. The talk will encompass topics such as intellectual property, team formation, financing/fundraising, and common pitfalls.

In short, this talk will serve as a window for aspiring young chemists to see how successful startups have commercialized chemical technologies. Additionally, investors will find this presentation useful as a robust database to sharpen their ability to recognize the patterns of success in chemical startups.

573571 Understanding Venture Capital Funding Decisions and Using Initial Funding to Build a Successful Startup

Many scientists and engineers do not have entrepreneurial experiences, and do not understand what investors look for in startups. The same scientific cruxes that are of interest in academia, often have no relevance to venture capitalists, investors of new businesses - also known as startups. In brevity, chemical engineering startups are no different than those of any other industries. Venture capitals invest in a firm's ability to grow, not profit, so that the firm can capitalize on their newly established market position sometime in the future. By the end of a firm's second year, venture capitals expect a handful of booked customers to show general viability of its product on the market; This is also an opportunity to conduct consumer research. In general, firms also need to provide concrete plans at this time to become profitable at some point in the future; though, these profit generalizations do not require actualization in the near or moderate future. Any further funding past this series-A round, depends on the firm's progress over their last two years under the venture capital's guidance. When a venture capital invests in a firm, they provide a substantial amount of advice and access to a vast business network, in addition to money. This two-year period of the startup is generally regarded as the trial and error stage on behalf of the venture capital. For example, most venture capitals fund dozens of startups for their initial two years, but fund only a couple of startups in their years after; though often at multiples in terms of dollar investments. In this presentation, we will go over how venture capitals decide which firms to fund, what firms should do in their first quarter, and present an outline of a standard 2-year plan.

581347 Chemical Engineering Innovations for a Renewable Economy

Chemical engineering evolved in the twentieth century with rise in the use of fossil resources and myriad of products derived from then. An array of these products required great innovations and made huge impact on the human race, and such examples include urea, transportation fuels, polymers, water purification chemicals, etc. Going forward as we make transition from fossil resources to renewable resources such as solar energy, wind and sustainably available biomass, we again have great opportunities for innovation that would not only make great impact but redefine chemical engineering. My talk will focus on my own experiences with challenges and opportunities presented to us as chemical engineers. I will share an overall perspective on the use of biomass as sustainable source of carbon, harvesting solar energy in conjunction with growing food without any land constraint, and creating solution processed solar cells that utilize chemical engineering principles along with knowledge from other engineering disciplines.

581350 Engineering Materials for Clinical Application

Dr. Schmidt's research has resulted in numerous patents that are licensed to companies, including AxoGen, Alafair Biosciences, Smith and Nephew, and Siluria Technologies. Comments for this panel will focus on peripheral nerve applications and one key success that is the foundation for the Avance Nerve Graft from AxoGen. In addition, along the way to uncovering strategies for nerve regeneration, Dr. Schmidt and her team also discovered a biomaterials processing approach that led to a separate application of protecting tendon's after surgery, resulting in the VersaWrap Tendon Protector from Alafair Biosciences, a start-up company in Austin, Texas.

Nerve Repair: Dr. Schmidt's technology for processing decellularized nerve tissue was licensed to AxoGen in Alachua, Florida (http://www.axogeninc.com) for the AVANCE^{TM} nerve graft. Many tens of thousands (~100,000) of AxoGen's AVANCE^{TM} grafts have been implanted in patients at hundreds of hospitals globally.

Post-surgical Wound Management: Using a novel in situ crystallization processing technique, Dr. Schmidt's team created mechanically robust hyaluronic acid-based hydrogel films that can be easily handled, sutured, and delivered laparoscopically. This research is the basis of Alafair Biosciences, a start-up company in Austin, TX (http://www.alafairbiosciences.com), which is run by Dr. Schmidt's PhD graduate and a neurosurgeon collaborator. Alafair's first product, VersaWrap^{TM} Tendon Protector, has been implanted in more than 500 patients since December 2017.

AUTHOR INDEX